About the Author . . .
Mary Powell has spent her adult life combining her math knowledge and business experience to offer a Unique Approach to teaching math.

Receiving "Outstanding Math Student" award in high school, she went on to college pursuing a degree in education with emphasis on mathematics. Due to a strange course of events she entered the business world long before the classroom.

Her business experience consists of Civil Engineering Draftsman for Exxon Petroleum; Business Office Supervisor for New York Bell Telephone; Licensed Real Estate Broker; President Po-Co Housing; Director Woodvine Townhomes, Inc.

She is author/publisher of MATHCO Educational Materials, which includes a Basic Math Program (classroom management program) and a Self-Tutoring Math KIT, both with all instruction on cassette tape in both English and Spanish; as well as two Practice Exercise Books and a line of Geometry Posters.

She has 12 years experience teaching 7, 8, and 9th grade math and 7 years operating her own tutorial school with emphasis on preparation for SAT, ACT, TAAS, TASP, in addition to tutoring recruits for the Armed Services Vocational Aptitude Battery (ASVAB Test).

She has conducted workshops across the country in "Training Teachers How To Teach Math Concepts". The last 6 years, prior to writing this book, were spent teaching math for Paris Junior College at the Greenville Technical Center in Greenville, Texas where she was nominated to be placed in Who's Who Among America's Teachers.

She is the proud mother of three grown sons and **Grandma Mary** to four beautiful grandchildren!

This book is dedicated to—
the many students that expressed
their appreciation for
"making math easier."

Table of Contents

MATHEMATICS AS A "LANGUAGE"

PARTS OF A MATH PROBLEM

In case you have forgotten, the parts of math problems have names just like the parts of your body, (and these are the words used on Standardized Tests).

Example: Find the product of 4 and the sum of 3 and 6. Did you get 36? Good for you!

$$4\ (3 + 6) = 4\ (9) = 36$$

Example: Find the quotient of 16 and the difference of 6 and 2.

$$16 \div (6 - 2) = 16 \div 4 = 4$$

All math problems can be written in EQUATION FORM and WORKING FORM.

Addition

EQUATION FORM	WORKING FORM
$4 + 5 = 9$	Addend 4 Addend + 5 Sum 9
Addend + Addend = Sum	

Since both parts of an addition problem have the same name, it doesn't matter which comes first. (Commutative Property of Addition)

Subtraction

EQUATION FORM	WORKING FORM
$8 - 5 = 3$ Minuend - Subtrahend = Difference	Minuend 8 Subtrahend -5 Difference 3

Since the parts of a subtraction problem do not have the same names, they must be written in the correct order. (Not Commutative)

Mutiplication

EQUATION FORM	WORKING FORM
$2 \times 5 = 10$ Factor X Factor = Product	Factor 5 Factor X2 Product 10

Again, both parts have the same name, so it doesn't matter which comes first. (Commutative Property of Multiplication)

DIVISION

EQUATION FORM

$6 \div 3 = 2$

Dividend ÷ Divisor = Quotient

WORKING FORM

$$\begin{array}{r} 2 \\ 3\overline{)6} \end{array}$$

2 Quotient; 6 Dividend; 3 Divisor

The parts have **different** names; so they must be written in the correct order. (Not Commutative)

READING MATH

\+ This is the only way to say "ADD" or "PLUS".

\- This is the only way to say "SUBTRACT" or "MINUS".

But look how many ways to say "MULTIPLY" or "TIMES".

X ( ) [ ] ab 2c

EXAMPLE:

$4 \times 3 = 12$ $\qquad$ $4(3) = 12$

$4 \cdot 3 = 12$ $\qquad$ $4[3] = 12$

Two letters side by side (ab) or a letter and a number side by side (2c) always say "MULTIPLY".

There are four ways to show "DIVISION".

÷ ⌐ ∟ ——

$12 \div 6 = 2$ $\qquad$ $\frac{12}{6} = 2$

$$\begin{array}{r} 2 \\ 6\overline{)12} \end{array} \qquad \begin{array}{l} 6\,\underline{|\,12} \\ \quad\ 2 \end{array}$$

FAMILY OF FACTS

Did you know you are twice as smart in subtraction as you are in addition? Every time you know one addition fact, you know **two** subtraction facts.

Example:

5 + 4 = 9	3 + 7 = 10
9 - 5 = 4	10 - 3 = 7
9 - 4 = 5	10 - 7 = 3

Also, you are twice as smart in division as you are in multiplication. Every time you know one multiplication fact, you know two division facts!

Example:

6 X 2 = 12	3 X 5 = 15
12 ÷ 6 = 2	15 ÷ 3 = 5
12 ÷ 2 = 6	15 ÷ 5 = 3

EQUALITY AND INEQUALITY

First graders learn to write the "equal mark," but do they know what it means? They might say "the answer," but it would be best if they said "is the same as."

Example:

$$\frac{16}{2} = 5 + 3$$

"Sixteen divided by 2 **is the same as** 5 plus 3." Why? Because $\frac{16}{2}$ is another way to write 8. "5 + 3" is another way to write 8. Therefore, **8 is the same as 8.**

If two quantities **are not equal,** then one is **greater than** or **less than** the other.

Example:

$<$ is read "less than"

$12 - 2 < 3(7)$

$10 < 21$

$>$ is read "greater than"

$20 + 9 > 7(4)$

$29 > 28$

NUMBER THEORY

Counting Numbers—1, 2, 3, 4, 5, . . . the numbers you learned to count.

Whole Numbers—0, 1, 2, 3, 4, 5, . . . the set of "counting numbers" **and** 0.

Even Whole Numbers—all Whole Numbers that end in 0, 2, 4, 6, or 8.

Odd Whole Numbers—all Whole Numbers that end in 1, 3, 5, 7, or 9.

Do you remember what a "factor" is? It is part of a multiplication problem. The "factors" of 8 are 1 X 8 and 2 X 4. Therefore, 8 has four "factors", 1, 2, 4, 8. Because of the relationship of multiplication and division,every number is divisible by it's factors.

$8 \div 1 = 8 \quad 8 \div 2 = 4 \quad 8 \div 4 = 2 \quad 8 \div 8 = 1$

PRIME NUMBERS

Any number greater than 1 that only has two factors (itself and 1) is said to be "Prime".

Example:

7 is a "Prime", because it only has two factors, 1 and 7.

9 is **not** a "Prime", because it has three factors, 1, 3, 9.

"2" is the only even number that is "Prime", because all other even numbers have 2 as a factor.

Therefore, the "Prime Numbers" between 1 and 100 are: 2, 3, 5, 7, 11, 13,17, 19,23, 29, 31, 37, 41, 43, 47, 53, 59, 61, 67, 71, 73, 79, 83, 89, 91, 97.

COMPOSITE NUMBERS

All numbers that are **NOT** "Prime" are said to be "Composite" Numbers.

FACTORS and GCF

Do you remember a "factor" is part of a multiplication problem? Also, remember a number is divisible by it's "factors".

Example:
To find the "factors" of 24, we want the numbers you multiply to get 24.

1 X 24 = 24   2 X 12 = 24   3 X 8 = 24   4 X 6 = 24; therefore, the "factors" of 24 are {1, 2, 3, 4, 6, 8, 12, 24} .

"Common Factors" of two or more numbers are the "factors" they have alike.

Example:
Find the "common factors" of 8 and 12.

Factors of 8 are {1, 2, 4, 8}
Factors of 12 are {1, 2, 3, 4, 6, 12}
"Common Factors " of 8 and 12 are {1, 2, 4}
Therefore, The **Greatest Common Factor** (G C F) of 8 and 12 is {4} .

FACT TO REMEMBER:

The "Greatest" Common Factor is usually a small number!!

The "Least" Common Multiple is usually a big number!!

The next page will cover this!!

MULTIPLES and LCM

What 's the difference in a "factor" and a "multiple" ?
"Multiples" are the products you get when you multiply a number times all the whole numbers.

Example:
Multiples of 4 are {0, 4, 8, 12, 16, 20, 24, 28 . . .}

Since "0" is a multiple of all whole numbers, we do not count it when calculating the LCM of two or more numbers. Also, please note the dots following 28. This indicates this set of numbers goes on and on! There is no end to the set of multiples!!

Example:
Find the "common multiples" of 8 and 12.

Multiples of 8 are {8, 16, 24, 32, 40, 48, 56, 64, 72 . . .}
Multiples of 12 are {12, 24, 36, 48, 60, 72, 84 . . .}
Common Multiples of 8 and 12 are {24, 48, 72 . . .}

Therefore, the **Least Common Multiple** (L C M) of 8 and 12 is { 24 }

FACT TO REMEMBER:

The "Greatest Common Factor" is usually a small number!!

The "Least Common Multiple" is usually a large number!!

ORDER OF OPERATION

Did you know you will get the wrong answer to a math problem unless you follow the steps in a certain order, called Order of Operation?

Have you heard the expression, "**P**lease **E**xcuse **M**y **D**ear **A**unt **S**ally"? This will help you remember the order in solving a problem.

P stands for parentheses, do the operation inside grouping symbols **first**.

E stands for exponents and this operation is done next.

M stands for multiplication.

D stands for division.

A stands for addition.

S stands for subtraction.

When there is multiplication and division, you must do whichever comes first, working from the left. When there is addition and subtraction, you must do whichever comes first, working from the left.

Examples:

$48 - 32 \div 2^3 =$

$48 - 32 \div 8 =$

$48 - 4 = 44$

$20 - 2(3)^2 =$

$20 - 2(9) =$

$20 - 18 = 2$

$6 \div [4 - (8-5)] + 2^2 =$

$6 \div [4 - 3] + 2^2 =$

$6 \div 1 + 4 =$

$6 + 4 = 10$

$24 - 18 \div 3 + 2 =$

$24 - 6 + 2 =$

$18 + 2 = 20$

DIVISIBILITY RULES

There are some "tricks" that help us look at numbers and quickly tell if certain numbers are "factors" of that number. (Remember a number is **divisible** by it's "factors.")

Rule of 2—If the number is an even number (ends in 0, 2, 4, 6, or 8.)

Rule of 3—If the sum of the digits is divisible by 3.

Rule of 4—If the last 2 digits is divisible by 4.

Rule of 5—If the number ends in a "5" or "0".

Rule of 6—If the number is divisible by 2 **and** 3.

Rule of 8—If the last 3 digits are divisible by 8.

Rule of 9—If the sum of the digits is divisible by 9.

Rule of 10—If the number ends in a "0".

Example:
Is 2 a factor of 786? Yes, because it ends in an even number, 6.

Is 3 a factor of 252? Yes, because the sum of the digits is divisible by 3.

2 + 5 + 2 = 9 } 3 will divide into 9, so it will divide into 252.

Example:
Try each of the above rules on this number: "49,320".

2 is a factor, because it ends in an even number.

3 is a factor, because the sum of the digits is 18 and 3 is a factor of 18.

4 is a factor, because 4 will divide into 20 (last two digits.)

5 is a factor, because it ends in a 0 or 5.

6 is a factor, because it is divisible by 2 **and** 3.

8 is a factor, because 8 will divide into 320 (last three digits).

9 is a factor, because the sum of the digits is 18 and 9 is a factor of 18.

10 is a factor, because it ends in a 0.

COMMON FRACTIONS
(Ugh !!)

Now come on, "Fractions" are not so bad! Sure, you have forgotten some of the steps, but let's get to it and learn them again! First thing, the vocabulary is important.

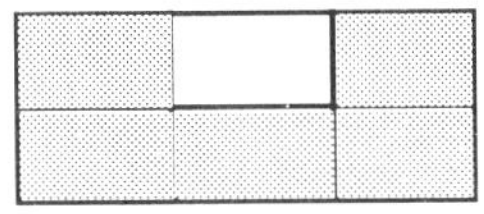

This rectangle is divided into 6 equal parts. 5 of those parts are shaded in; therefore, this is a picture of the Fraction $\frac{5}{6}$. The Denominator (bottom number) tells the total number of parts. The Numerator (top number) tells how many of those parts we are talking about. The Horizontal Bar is really another way to say "divide".

$\frac{5}{7}$ is a Proper Fraction (big number on bottom—in the denominator).

$\frac{8}{3}$ is an Improper Fraction (big number on top—in the numerator).

$2\frac{1}{4}$ is a Mixed Number (mixture of whole number and proper fraction).

How to Change an Improper Fraction to a Mixed Number.

- Divide the numerator by the denominator and make a fraction by placing the remainder over the denominator (or divisor):

$$\frac{10}{7} = 1\frac{3}{7} \qquad \left(7\overline{)10} = 1\frac{3}{7};\ 10 - 7 = 3 \right)$$

How to Change a Mixed Number to an Improper Fraction.

- Multiply the denominator times the whole number; add the numerator and place this number over the denominator:

$$2\frac{3}{5} = \frac{13}{5} \qquad \left(\frac{(5 \times 2) + 3}{5}\right) = \frac{10 + 3}{5} = \frac{13}{5}$$

MULTIPLICATION OF FRACTIONS

Multiply numerator times numerator and denominator times denominator .

$$\frac{2}{7} \times \frac{1}{9} = \frac{2}{63}$$

If you have a mixed number or a whole number, these must be changed to fractions first. (Remember, you can't multiply two fractions unless you have two fractions!!)

$$1\frac{3}{5} \times 4 =$$

$$\frac{8}{5} \times \frac{4}{1} = \frac{32}{5} = 6\frac{2}{5}$$

DIVISION OF FRACTIONS

First, we need to learn what a Reciprocal is. When two numbers are multiplied together and the answer is "1"; then they are said to be Reciprocals.

The reciprocal of $\frac{3}{4}$ is $\frac{4}{3}$. The reciprocal of $2\frac{1}{4}$ is $\frac{4}{9}$.

$$\left(\begin{array}{l} 2\frac{1}{4} = \frac{9}{4} \\ \frac{9}{4} \times \frac{4}{9} = \frac{36}{36} = 1 \end{array} \right)$$

Get it? Now you can say "turn it upside down" as long as you really know what a Reciprocal is!

Rule for Division of Fractions:
Change the problem to a multiplication problem by multiplying the dividend (first number) by the reciprocal of the divisor (second number). Then follow the rules for **Multiplication of Fractions**.

$$\frac{5}{8} \div \frac{2}{3} = \qquad\qquad 2\frac{1}{3} \div 8 = \frac{7}{3} \div \frac{8}{1} =$$

$$\frac{5}{8} \times \frac{3}{2} = \frac{15}{16} \qquad\qquad \frac{7}{3} \times \frac{1}{8} = \frac{7}{24}$$

ADDITION AND SUBTRACTION OF FRACTIONS WITH LIKE DENOMINATORS

Rule;
If the denominators are the same (alike); then just add or subtract the numerators and keep the common denominator (same).

$$\frac{1}{7} + \frac{5}{7} = \frac{6}{7} \qquad\qquad \frac{7}{9} - \frac{5}{9} = \frac{2}{9}$$

$$\begin{array}{r} 3\frac{1}{9} \\ 4\frac{2}{9} \\ +\ 1\frac{1}{9} \\ \hline 8\frac{4}{9} \end{array} \qquad\qquad \begin{array}{r} 12\frac{5}{13} \\ -\ 4\frac{4}{13} \\ \hline 8\frac{1}{13} \end{array}$$

ADDITION AND SUBTRACTION OF FRACTIONS WITH UNLIKE DENOMINATORS

To add or subtract fractions, they must have the **same denominators**.

However, to multiply and divide they can have **different denominators**.

Now go back and refresh your memory about **Least Common Multiples (LCM)**, because we are going to talk about **Least Common Denominators (LCD)**!!

"The LCD of two or more fractions is the LCM of their denominators."

Example:

$$\frac{2}{3} \quad \frac{1}{4} \quad \frac{5}{6}$$

The denominators of these three fractions are 3, 4, and 6. The LCM is 12; therefore, the **Least Common Denominator (LCD)** of these fractions is 12.

Rule to add and subtract fractions with **Unlike Denominators**:

Rename the fractions to have the same denominators. The LCD tells us what the new denominator should be.

When we rename a fraction to have a different denominator,this is referred to as "changing a fraction to an **equivalent fraction**." Do you realize that we do not change the value of any number when we multiply or divide by **one**?

Now look at how many ways one can be written:

$$\frac{5}{5} \quad \frac{12}{12} \quad \frac{43}{43}$$

Get it? So change $\frac{3}{4}$ to a fraction with a denominator of 12. All we do is multiply by **one**!!

$$\frac{3}{4} \cdot \underbrace{\left(\frac{3}{3}\right)}_{\text{ONE}} = \frac{9}{12} \qquad \frac{1}{2} \cdot \underbrace{\left(\frac{6}{6}\right)}_{\text{ONE}} = \frac{6}{12} \qquad \frac{5}{9} \cdot \underbrace{\left(\frac{2}{2}\right)}_{\text{ONE}} = \frac{10}{18}$$

After the unlike fractions have been renamed to have the same denominators, just follow the rules for **Addition and Subtraction with Like Denominators**. Study the examples on the next page carefully.

Examples to Study for Addition and Subtraction of Fractions with Unlike Denominators

Examples of Addition:

$$\begin{array}{r} 1\frac{3}{4}\left(\frac{3}{3}\right) = 1\frac{9}{12} \\ +5\frac{1}{6}\left(\frac{2}{2}\right) = 5\frac{2}{12} \\ \hline 6\frac{11}{12} \end{array}$$

$$\begin{array}{r} 3\frac{4}{9}\left(\frac{2}{2}\right) = 3\frac{8}{18} \\ +4\frac{5}{6}\left(\frac{3}{3}\right) = 4\frac{15}{18} \\ \hline 7\frac{23}{18} \end{array} = 7 + 1 + \frac{5}{18} = 8\frac{5}{18}$$

Examples of Subtraction:

$$\begin{array}{r} 8\frac{3}{4} \\ -3 \\ \hline 5\frac{3}{4} \end{array}$$

$$\begin{array}{r} 7 = 6\frac{5}{5} \\ -4\frac{3}{5} = 4\frac{3}{5} \\ \hline 2\frac{2}{5} \end{array}$$

$$\begin{array}{r} 5\frac{1}{3}\left(\frac{3}{3}\right) = 5\frac{3}{9} = 4 + \frac{9}{9} + \frac{3}{9} = 4\frac{12}{9} \\ -3\frac{5}{9}\left(\frac{1}{1}\right) = 3\frac{5}{9} = \qquad = 3\frac{5}{9} \\ \hline 1\frac{7}{9} \end{array}$$

Reducing Fractions

The only time a fraction needs to be reduced is when the numerator and denominator share a "common factor"!! Remember how we **renamed** a fraction by multiplying by **one** $\left[\frac{5}{7}\left(\frac{2}{2}\right) = \frac{10}{14}\right]$ We used the LCM (least common multiple). Now, to **reduce a fraction** we are going to use the GCF (greatest common factor). In the fraction $\frac{18}{24}$, we know it can be reduced because the numerator of 18 and the denominator of 24 share several common factors (2, 3, 6). The **greatest common factor** of 18 and 24 is 6. Next, we divide the numerator and denominator by 6. You could have used **any** common factor, but you get through quicker if you use the **greatest**!

$$\frac{18 \div (6) = 3}{24 \div (6) = 4} \qquad \frac{4 \div (4) = 1}{12 \div (4) = 3}$$

$$\frac{24 \div (12) = 2}{60 \div (12) = 5} \qquad \frac{18 \div (9) = 2 \quad = 2 \div (2) = 1}{54 \div (9) = 6 \quad = 6 \div (2) = 3}$$

POWERS OF TEN

Looking at this number, 2^3, the 2 is called the base and the 3 is called the exponent. An exponent (or power) tells us how many times to use the base as a factor (part of a multiplication problem).

$10^1 = 10 = 10$

$10^2 = 10 \times 10 = 100$

$10^3 = 10 \times 10 \times 10 = 1000$

$10^4 = 10 \times 10 \times 10 \times 10 = 10000$

$10^5 = 10 \times 10 \times 10 \times 10 \times 10 = 100000$

When we multiply or divide by 10, 100, 1000 or any other **Power of Ten**. we can use a short-cut method:

"To multiply by a **POWER OF TEN** just move the decimal point to the **right** as many places as the exponent (or the number of zeros)".

Example:

$3.14 \times 10 = 3.14 \times 10^1 = 31.4$

$.672 \times 100 = .672 \times 10^2 = 67.2$

"To divide by a **POWER OF TEN** just move the decimal point to the **left** as many places as the exponent (or the number of zeros)".

Example:

$15.6 \div 10 = 15.6 \div 10^1 = 1.56$

$300.74 \div 100 = 300.74 \div 10^2 = 3.0074$

$32 \div 1000 = 32 \div 10^3 = .032$

DECIMAL FRACTIONS

You have studied Common Fractions, now we will study Decimal Fractions. Please do **not** say "**POINT**" in reading a decimal. Give it correct place value.

Examples: Both of these numbers are "fifteen **and** six hundredths"

$15\frac{6}{100}$ 15.06

Both of these numbers are "twelve **and** nine tenths"

$12\frac{9}{10}$ 12.9

Never say "and" in reading a number, except when there is a decimal point! By the way, how much money is this? **.99¢** If you said, "Ninety-nine cents," you are wrong! It is **Ninety-Nine Hundredths of a Cent!!** All digits on the left of the decimal are whole numbers. All numbers on the **right** of the decimal are fractions less than one! "Ninety-nine cents" can be written $.99 **Ninety-Nine Hundredths of a Dollar!!** or 99¢. (Boy, have we been hearing a lot of wrong stuff!)

(Addition and Subtraction of Decimal)

Keep the decimal points lined up. (If a number does not have a decimal point, it is a **whole number** and goes on the left of the decimal point.) **Fractions** are on the right. **Mixed numbers** are on the left and right.

Keep the decimal points lined up; fill in all blank spaces with zeros then add or subtract as in whole numbers.

Example:

3.05 + 26.009 = 14 - .685 =

```
  03.050        14.000
 +26.009       -00.685
  29.059        13.315
```

MULTIPLICATION OF DECIMALS

Just multiply as in whole numbers; then count the number of digits on the right of the decimal point in both factors and add them together. Count off that many digits in the product, from the right.

```
   .64 (2)           17.3 (1)
  X3.2 (1)          X.005 (3)
   128              .0865 (4)
  192
 2.048 (3)
```

DIVISION OF DECIMALS

To divide decimals, the divisor must be a whole number. If it is, place the decimal point in the quotient directly above the decimal point in the dividend. Then divide as in whole numbers.

Example:

```
    43.59
 4 |174.36
   16
    14
    12
     23
     20
      36
      36
```

If the divisor if **not** a whole number multiply the **divisor and dividend** by a **Power of Ten** to make it a whole number. (Remember all we do in multiplying and dividing by **Powers of Ten** is move a decimal point!) Then place the decimal point in the **quotient** directly above the decimal point in the **dividend**. Continue by dividing as in whole numbers.

Example:

```
      13800.
 .05 |690.00
      5
      19
      15
       40
       40
        00
         0
         00
          0
```

```
         5000.
 .007 |35.000
       35
        00
         0
         00
          0
          00
           0
```

PERCENT FRACTIONS

Are you surprised to see the word Fraction used with Percent ? Well, that is correct. We have three kinds of Fractions: Common Fractions, Decimal Fractions and Percent Fractions! Remember, we said earlier "do not say **point**."

Use correct language in reading these fractions. The word **Percent** and the word **Hundredths** are interchangable words.

$.17$	=	$\frac{17}{100}$	=	17%
17 Hundredths		17 Hundredths		17 Percent

Remember how to multiply and divide by **Powers of Ten** by moving the decimal point to the right when we multiply and to the left when we divide. Now look at these examples.

To change a Percent to a Decimal: Move the decimal point 2 places to the **left** and drop the percent sign.

$87\% = .87$ $125\% = 1.25$ $1\% = .01$

To change a Decimal to a Percent: Move the decimal point 2 places to the **right** and add the percent sign.

$.73 = 73\%$ $.02 = 2\%$ $2.07 = 207\%$

To change a Percent to a Fraction: Drop the percent sign. Place the number over 100 and simplify.

$$16\% = \frac{16}{100} = \frac{4}{25}$$

To change a Fraction to a Percent: Change the fraction to a decimal (divide the denominator into the numerator). Then follow the rules for changing a decimal to a percent.

$$\frac{4}{5} = 5\overline{)4.0}\ \text{quotient } .8,\ \text{subtract } 40 \quad \rightarrow \quad .8 = 80\%$$

SOLVING PERCENT PROBLEMS

All Percent Problems can be solved by one formula. There are 3 parts to every percent problem__ 2 of those parts must be given in order to find the missing part. (This means reading comprehension!)
No matter how the problem is worded, **re-write** the sentence in this form!

R B P

__________% of __________________ is ______________

"R" stands for **RATE** and is **always** in front of the % sign.

"B" stands for **BASE** and is **always** after the word "of".

"P" stands for **PERCENTAGE** and is aways with the word "is",(either before or after, and remember " = " is read "is").

After filling in these blanks, then use this proportion to solve for the missing part.

$$\frac{R}{100} = \frac{P}{B}$$

Study these examples:

1) 18 is what percent of 54 ? __X__% of **54** is **18** . Find the RATE.

2) 29% of what number is 35? **29%** of __X__ is **35**. Find the BASE.

3) 6% of 450 is ____? **6%** of **450** is __X__? Find the PERCENTAGE.

We will have more problems in the section on "Word Problems".

1.

$$\frac{X}{100} = \frac{18}{54}$$

$$54\ X = 18\ (100)$$
$$54\ X = 1800$$
$$\frac{54\ X}{54} = \frac{1800}{54}$$

$$X = 33\frac{1}{3}$$

2.

$$\frac{29}{100} = \frac{35}{X}$$

$$29\ X = 35\ (100)$$
$$29\ X = 3500$$
$$\frac{29\ X}{29} = \frac{3500}{29}$$

$$X = 120.69$$

3.

$$\frac{6}{100} = \frac{X}{450}$$

$$100\ X = 6\ (450)$$
$$100\ X = 2700$$
$$\frac{100\ X}{100} = \frac{2700}{100}$$

$$X = 27$$

VOLCABULARY OF GEOMETY

POINT—An "idea" represented by a dot, and indentified by a capital letter.

LINE— A series of points (very close together) that has no beginning and no end. It is identified by 2 points

$\overline{AB}$ is read, "Line AB"

LINE SEGMENT— "Part" of a line that has a beginning and an end point.

$\overline{CD}$ is read, "Line segment CD"

RAY—"Part" of a line that has a beginning and no end point.

$\overrightarrow{RS}$ is read ,"Ray RS".

ANGLE—Two rays that share a common end point.

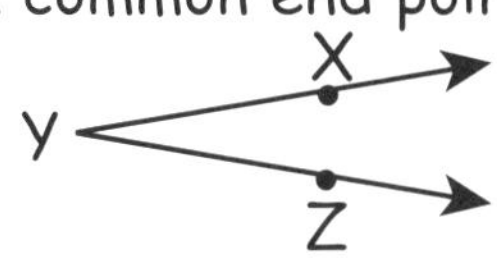

$\angle$ XYZ is read "Angle XYZ".

INTERSECTING LINES— When lines cross at a common point, they are "intersecting lines."

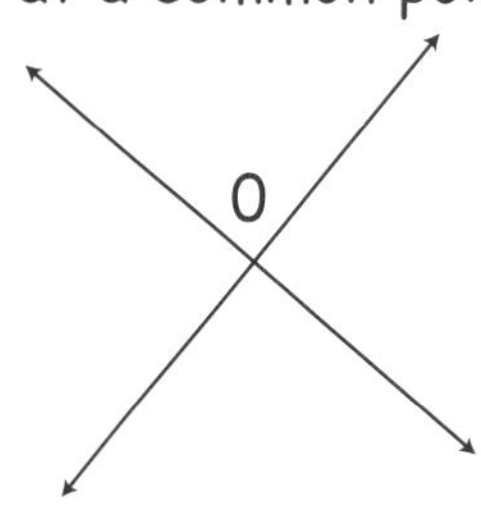

PARALLEL LINES— Lines that are the same distance apart at all points are "parallel lines". (They will never intersect.)

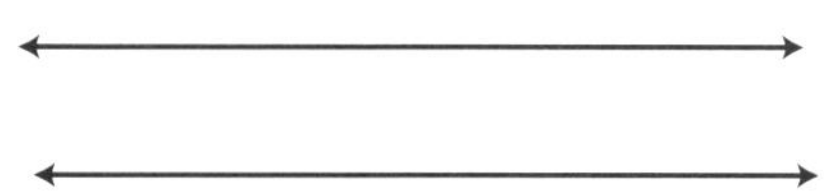

PERPENDICULAR LINES—Lines that intersect at right angles (90°) are said to be "perpendicular."

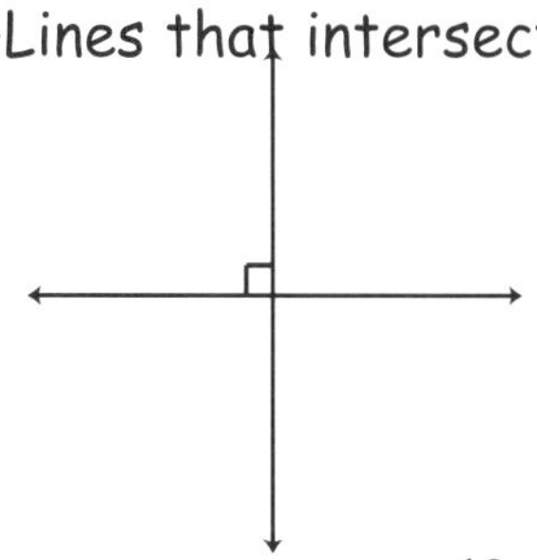

POLYGONS

A "Polygon" is a geometric figure that is completely enclosed by Line segments.(no curves) Look at this chart which shows some familiar polygons. The Highway Dept. uses Poygons for their road signs. Long before a child can read he can recognize certain shapes.

TRIANGLE—3 sided polygon **QUADRILATERAL**—4 sided polygon

PENTAGON— 5 sided polygon **HEXAGON**—6 sided polygon

OCTAGON—8 sided polygon

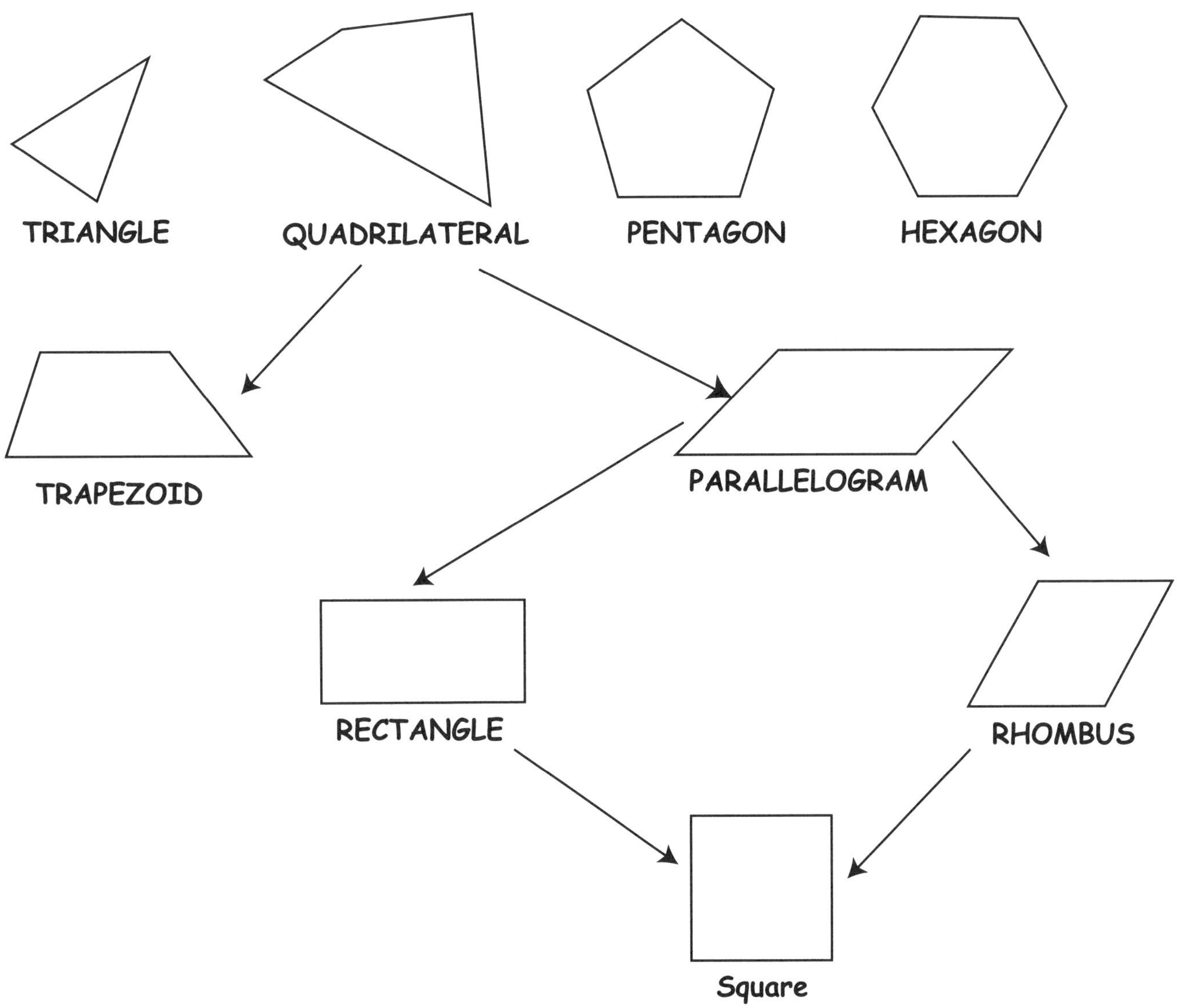

QUADRILATERALS

Continuing our study of "Polygons", let us take just the **Quadrilateral** and see how many different 4-sided polygons there are. Look at the special characteristics of each:

Trapezoid—a quadrilateral with 1 pair of parallel sides.

Parallelogram—a quadrilateral with 2 pair of parallel sides.

Rectangle—a parallelogram (a quadrilateral) with 4 right angles (90^0)

Rhombus—a parallelogram (a quadrilateral) with all sides equal in length.

Square—a rectangle (a quadrilateral) with all sides equal in length or a rhombus with 4 right angles.

Try these questions

	ANSWER
1. Is every rectangle a parallelogram?	yes
2. Is every quadrilateral a rectangle ?	no
3. Is every rectangle a quadrilateral ?	yes
4. Is every square a rhombus ?	yes
5. Is every rhombus a square ?	no

Think about this:

A square is a rectangle, is a parallelogram, is a rhombus, is a quadrilateral, is a polygon!!!

TRIANGLES

A **Triangle** is a 3-sided polygon. It also has 3 angles.

"The sum of the measure of the angles of ALL triangles add up to 180^0."

Look at these triangles and see if you can find how many degrees are in the missing angle.

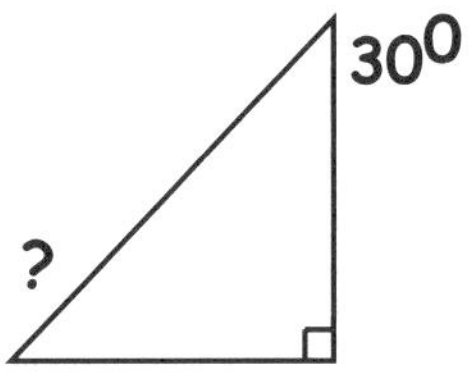

Answer: 60^0

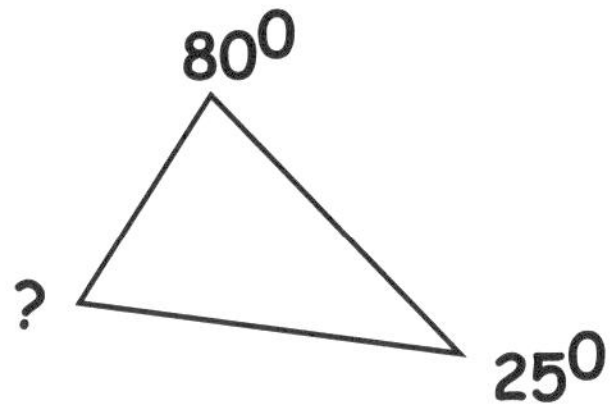

Answer:75^0

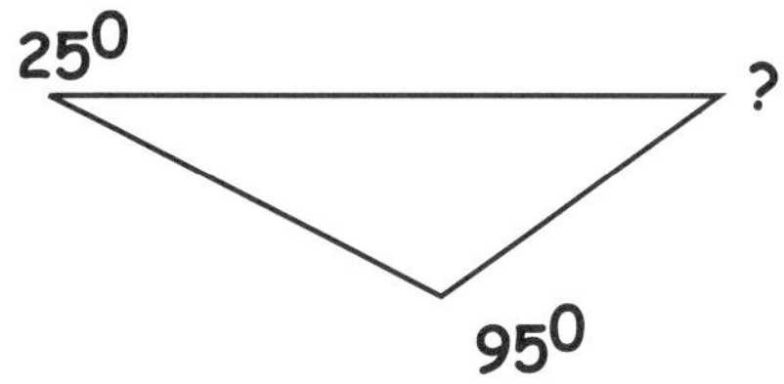

Answer:60^0

(Identifying Triangles)

SCALENE
(No sides same length)

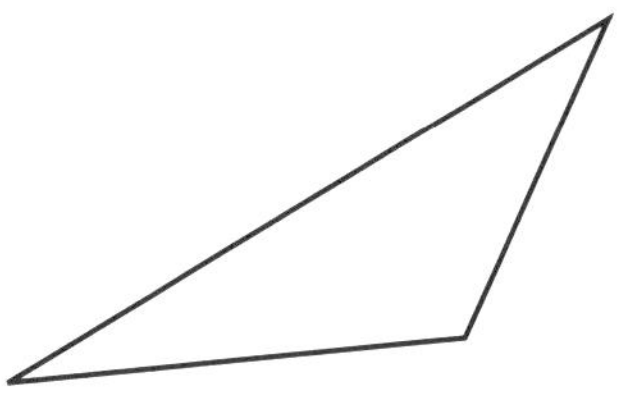

ISOSCELES
(Two sides same length)

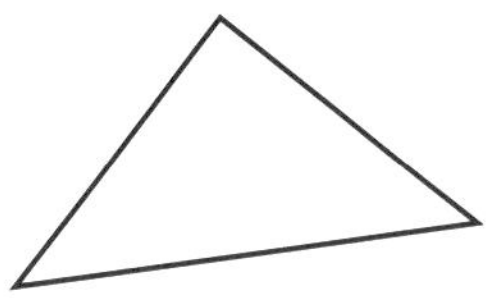

EQUILATERAL
(All 3 sides same length)

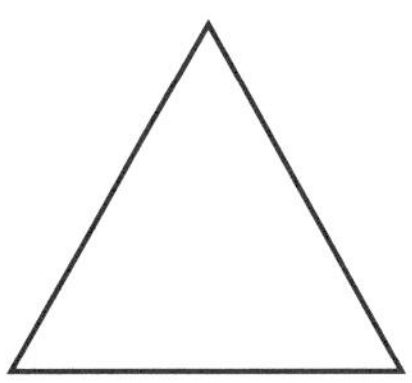

RIGHT
(Triangle with Right Angle)

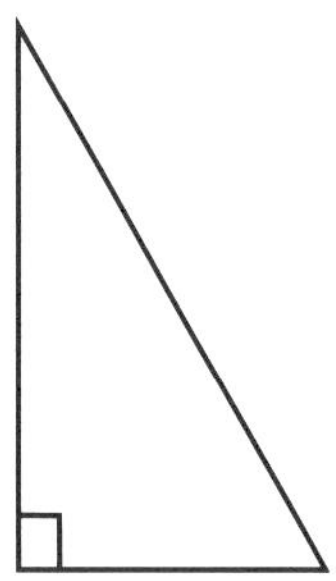

ANGLES

Going back to the page covering Vocabulary of Geometry, you 'll remember that an angle is made by "two rays that share a common endpoint." In reading an angle such as $\angle$CDF, the middle letter has to be the **vertex** of the angle, (the point the two rays share in common). We measure angles by degrees.

There are 360^0 in every circle. Half a circle is a Straight Line **which is also called a** Straight Angle; therefore, half of 360^0 is 180^0.

The instrument used to measure angles is called a Protractor.

Examples:

Acute Angle (less than 90^0)

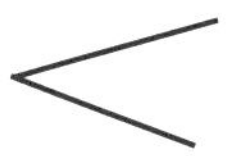

Obtuse Angle (more than 90^0 but less than 180^0)

Straight Angle (exactly 180^0)

Right Angle (exactly 90^0)

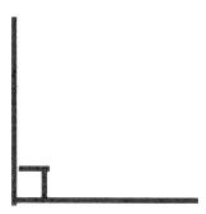

Complementary Angles (2 or more angles = 90^0)

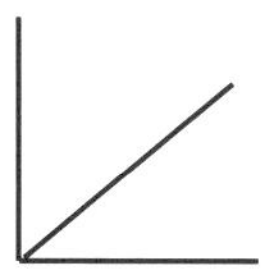

Supplementary Angles (2 or more anlges = 180^0)

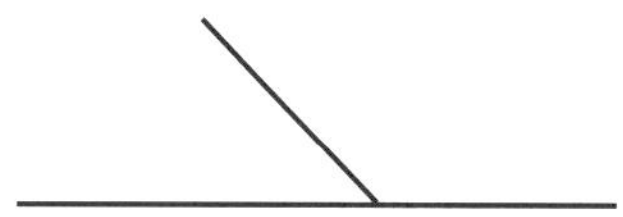

THE SUM OF THE ANGLES Of EVERY TRIANGLE is 180^0.

In this figure , two parallel lines are cut by a transversal forming 8 angles.

If < a = 120^0; then < b= 60^0 (Supplementary). Likewise < a and < c; < c and < d.

We just proved < a = < d and < b= < c } **VERTICLE ANGLES**

< a = < e } **CORRESPONDING ANGLES**. Just as in the upper intersection, we can go around and prove the measure of all 8 angles!

< c and < f ; < d and < e are **ALTERNATE INTERIOR ANGLES.**

< a and < h; < b and < g are **ALTERNATE EXTERIOR ANGLES.**

Nearly all Standardized Tests require the information on this page to be a part of your vocabulary!!

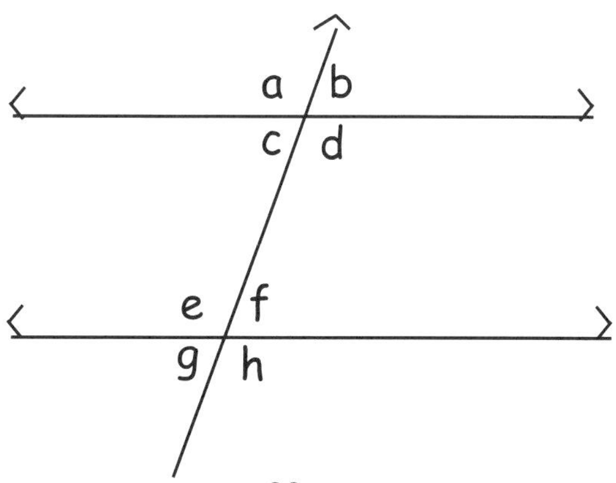

PERIMETER AND AREA OF POLYGONS

Remeber that a Polygon is a geometeric figure completely enclosed by **line segments** (no curves!). To find the **perimeter** of any polygon, we are going to add the lengths of all the sides! No big deal? To find the **area** of any polygon, we are going to find how many **square units** are inside the figure. However, we do have **formulas** for certain geometric figures that just make it easier to follow.

(A **formula** is just like a **recipe** in cooking—it tells all the ingredients and what to do wth each!!) Therefore, **always write the formula first!** Look at the examples below for the Rectangle, Square and Triangle. On the left of each figure is the formula for Area and how to follow it. On the right is the formula for Perimeter and how to follow it. **Sometimes the measure of each part is not given, but enough information is given for you to find what you need!**

Area		**Perimeter**

Rectangle

Area

$A = lw$

$A = (4)(3)$

$A = 12$ sq. ft.

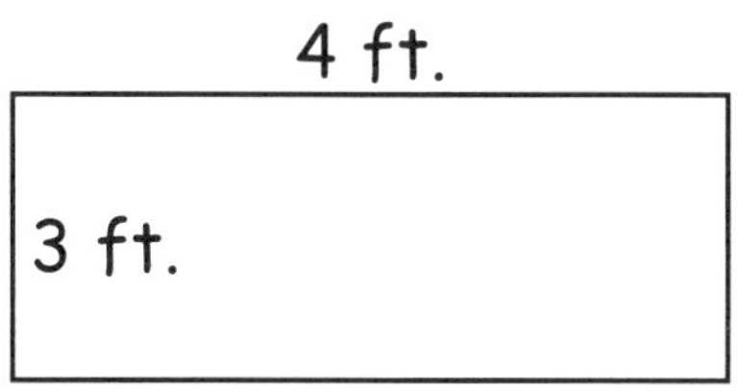

Perimeter

$P = 2L + 2W$

$P = 2(4) + 2(3)$

$P = 8 + 6$

$P = 14$ ft.

Square

Area

$A = s^2$

$A = (5^2)$

$A = (5)(5)$

$A = 25$ sq. in.

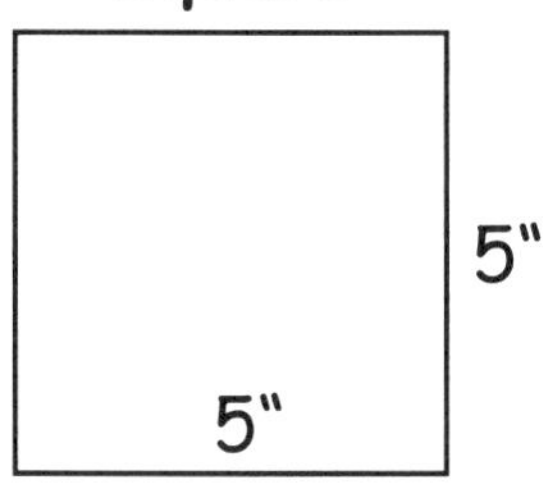

Perimeter

$P = 4S$

$P = 4(5)$

$P = 20$ in.

Triangle

Area

$A = \frac{bh}{2}$

$A = \frac{(8)(4)}{2}$

$A = \frac{32}{2}$

$A = 16$ sq. ft.

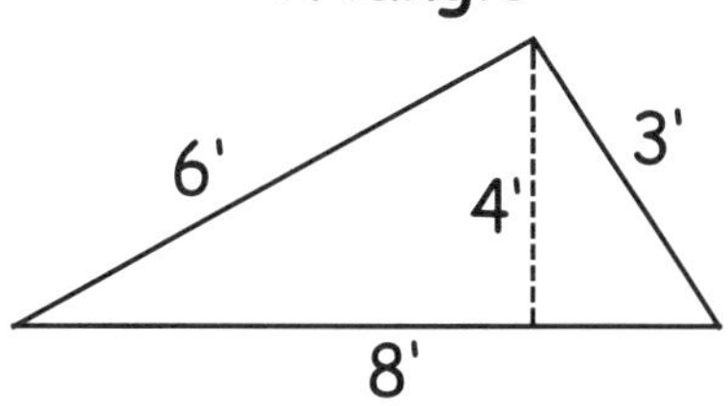

Perimeter

$P = a + b + c$

$P = 3 + 6 + 8$

$P = 9 + 8$

$P = 17$ ft.

PYTHAGOREAN THEOREM

SQUARE AND SQUARE ROOTS

In every Right Triangle a special relationship exists and this is called the **Pythagorean Theorem**, (named for a Greek mathematician). Look at the example as you read this:

In every right triangle, the square of the hypotenuse is always equal to the sum of the squares of the 2 legs

$c^2 = a^2 + b^2$
$c^2 = (3)^2 + (4)^2$
$c2 = 9 + 16$
$\sqrt{c^2} = \sqrt{25}$
$c = 5$

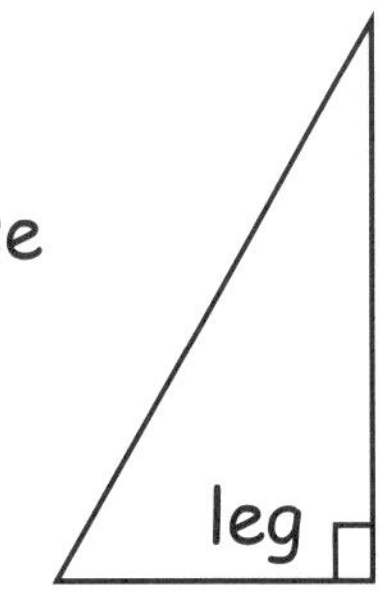

EXAMPLES:

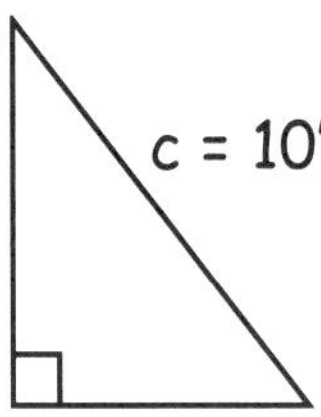

$c^2 = a^2 + b^2$
$(10)^2 = a^2 + (8)^2$
$100 = a^2 + 64$
$100-64 = a^2 + 64 - 64$
$36 = a^2$
$\sqrt{36} = \sqrt{a^2}$
$6 = a$

(Squares and Square Roots)

Refresh your memory about Perfect Squares. $3^2 = 9$; $5^2 = 25$; $9^2 = 81$
$10^2 = 100$; $11^2 = 121$; $12^2 = 144$; $13^2 = 169$ . . . The Square of a number is the product when a number is "multiplied by itself."

Now, the **Square Root** is the "number that was multiplied by itself"!!

$\sqrt{25} = 5$; $\sqrt{81} = 9$; $\sqrt{144} = 12$; $\sqrt{169} = 13$; $\sqrt{100} = 10$ . . .

CIRCLES

A circle is a set of points all equal distance from a point called the **center** of the circle.

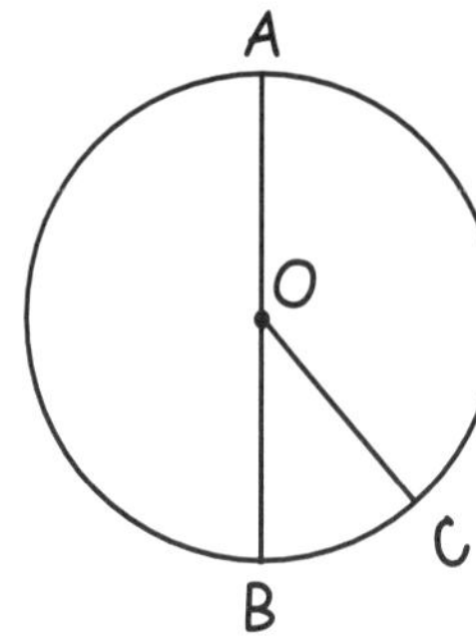

Looking at the picture:
A **RADIUS** is a line segment from one point on the circle to the center. $\overline{Ao}, \overline{Bo}, \overline{Co}$
A **DIAMETER** is a line segment from one point on the circle to another point on the circle, which passes through the center. $\overline{AOB}$

A **RADIUS** is half a **DIAMETER**!
A **DIAMETER** is 2 times the **RADIUS**!

The **CIRCUMFERENCE** is the distance around a circle.
The **AREA** is the number of square units of measure inside the circle.

Important

In **every** circle, **If the circumference is divided by the diameter the answer will be 3.14.)** (Actually, this is a non-repeating decimal which goes on and on.) This number is called pi and is identified by this Greek letter π. The fractional value of π is $\frac{22}{7}$.

The formula for the **Circumference** is $C = \pi d$
The formula for the **Area** is $A = \pi r^2$

EXAMPLES:

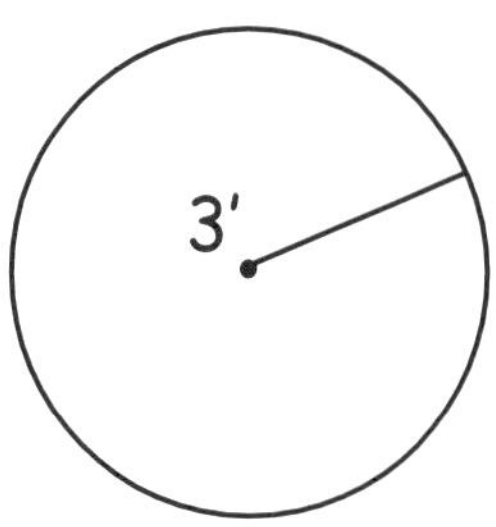	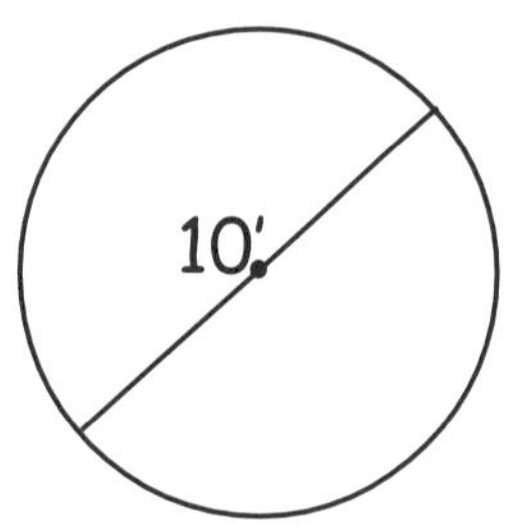	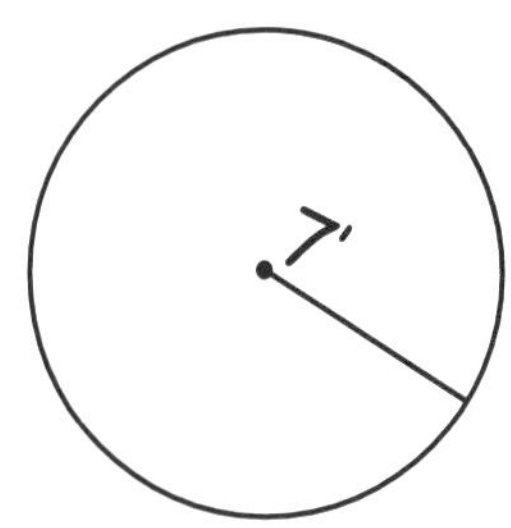
$C = \pi d$ $C = (3.14)\ (6)$ $C = 18.84$ ft.	$C = \pi d$ $C = (3.14)\ (10)$ $C = 31.4$ ft.	$C = \pi d$ $C = \left(\frac{22}{7}\right)\left(\frac{14}{1}\right)$ $C = \frac{308}{7}$ $C = 4'4$ ft.
$A = \pi r^2$ $A = (3.14)\ (3)^2$ $A = (3.14)\ (9)$ $A = 28.26$ sq. ft.	$A = \pi r^2$ $A = (3.14)\ (5)^2$ $A = (3.14)\ (25)$ $A = 78.5$ sq. ft.	$C = \pi r^2$ $A = \left(\frac{22}{7}\right)\ (7)^2$ $C = \left(\frac{22}{\not{7}_1}\right)\left(\frac{\not{49}^7}{1}\right)$ $C = 154$ sq. ft.

INTEGERS

As you already know, we have Positive and Negative Numbers called INTEGERS.

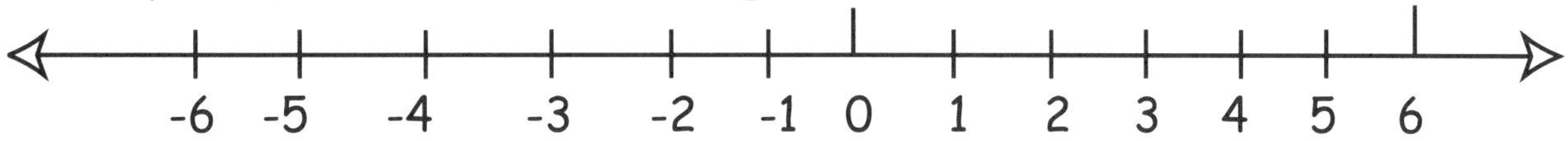

Positive Integers are on the **right** of "0" on the number line.

Negative Integers are on the **left** of "0" on the number line.

For every number on the right (whole numbers), it's Opposite is on the left. When Opposites are added, the answer is "0". As in Whole Numbers, Fractions, and Decimals there are rules for addition, subtraction, multiplication and division.

Multiplication and Division (same rule)

If the signs **are alike**, just multiply or divide and the answer is positive.
If the signs **are not alike**, just multiply or divide and the answer is negative.

5 X 3 = 15
(-4) X (-2) = 8
6 X (-3) = -18

12 ÷ 3 = 4
(-6) ÷ (-2) = 3
12 ÷ (-2) = -6

Addition

If the signs are **alike**, add and take the common sign.

5 + 3 = 8
(-6) + (-1) = -7

9 + 2 = 11
(-3) + (-7) = -10

If the signs **are not alike**, take the difference (remember that means subtract) and take the sign of the larger .

5 + (-2) = 3
(-8) + 6 = -2

(-16) + 18 = 2
9 + (-12) = -3

Subtraction

Actually, there is no such thing as subtraction__ we only add the opposite! So when you see an "old fashioned subtract sign", change the problem to an addition problem by adding the opposite of the subtrahend (second number). Then follow the rules for addition.

6 - 3 = 6 + (-3) = 3
(-12) - (-8) = (-12) + 8 = -4
20 - (-1) = 20 + 1 = 21
(-5) - (-2) = (-5) + 2 = -3
13 - 20 = 13 + (-20) = -7

EQUATIONS

An EQUATION is a mathematical sentence stating everything on the left of the equal mark **is the same as** everything on the **right**.

$$6 + 4 = 5(2) \qquad 9 - 3 = 12 - 6 \qquad 2(14) = 4(7)$$

Now what happens when one of the numbers is missing? We let a letter (called a variable) hold the place for the missing number. Looking at the examples below, you probably will agree you need a **method** to solve all equations. The object is to get the variable on one side of the equal mark and a number on the other side! However, you must follow mathematical laws for **everything** that is done!

EXAMPLES:

$-5n + 8 = -12$

$-5n + 8 - 8 = -8 - 12$ ← Add -8 to both sides

$-5n = -20$ ← Simplify

$\frac{-5n}{-5} = \frac{-20}{-5}$ ← Divide both sides by -5

$n = 4$ ← Simplify

Check

$-5n + 8 = -12$

$-5(4) + 8 = -12$

$-20 + 8 = -12$

$-12 = -12$

$5X - 6 = 4X + 9$

$5X - 4X - 6 = -4X + 4X + 9$ ← Add -4 X to both sides

$X - 6 = 9$ ← Simplify

$X - 6 + 6 = 9 + 6$ Add +6 to both sides

$X = 15$ ← Simplify

Check

$5X - 6 = 4X + 9$

$5(15) - 6 = 4(15) + 9$

$75 - 6 = 60 + 9$

$69 = 69$

The CHECK is your PROOF! Plug in your answer to see if the left **is the same as** the right!!

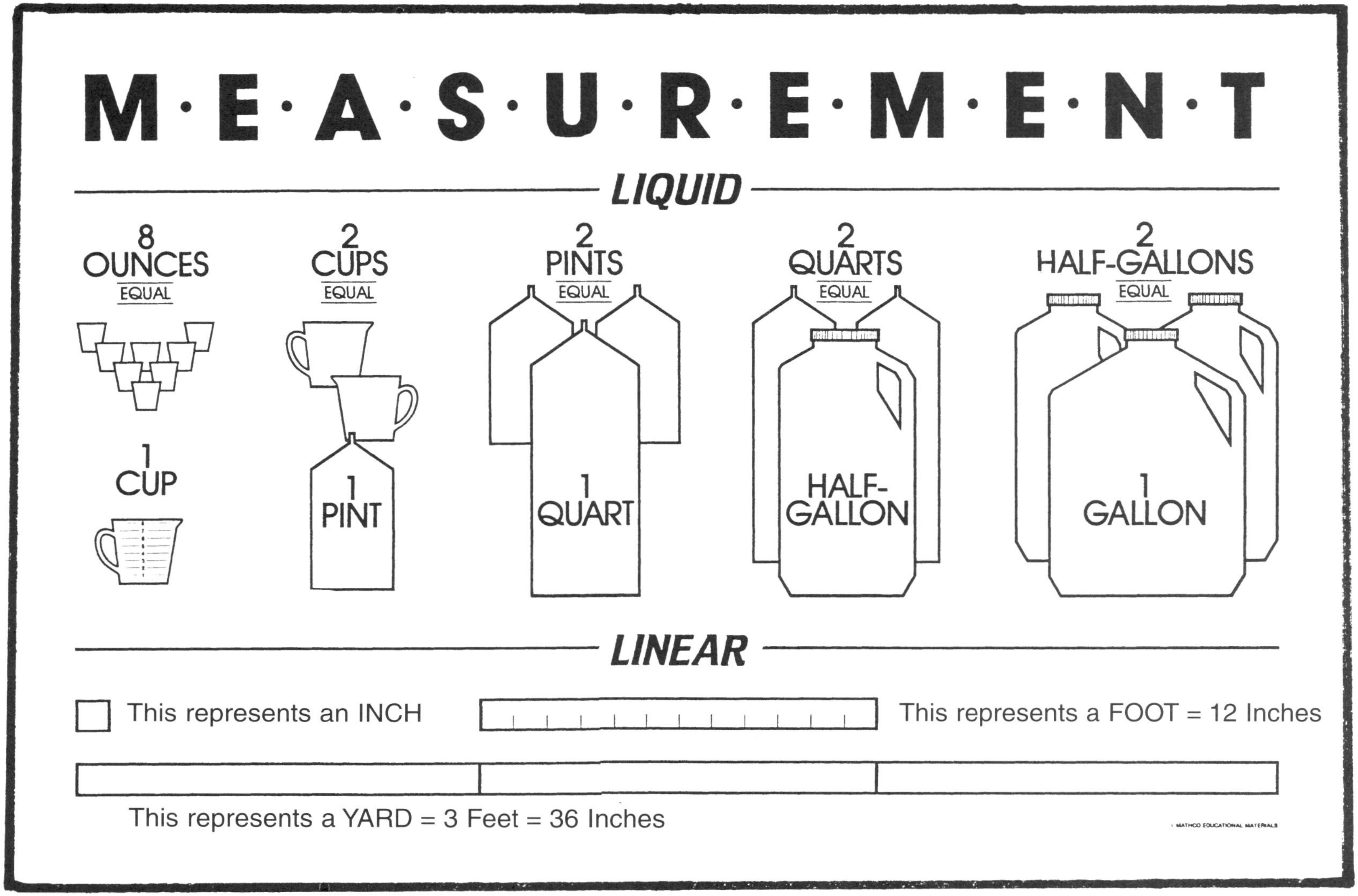
M·E·A·S·U·R·E·M·E·N·T
LIQUID
8 OUNCES
EQUAL
1 CUP
2 CUPS
EQUAL
1 PINT
2 PINTS
EQUAL
1 QUART
2 QUARTS
EQUAL
HALF-GALLON
2 HALF-GALLONS
EQUAL
1 GALLON
LINEAR
This represents an INCH
This represents a FOOT = 12 Inches
This represents a YARD = 3 Feet = 36 Inches
MATHCO EDUCATIONAL MATERIALS

MEASUREMENT

(Customary Units Of Measure)
On the previous page you'll see a graphic description of **Liquid Measure**. You'll see 8 liquid ounces = 1 cup; 2 cups = 1 pint; 2 pints = 1 quart; 2 quarts = 1 half-gallon; 2 half-gallons = 1 gallon. Isn't that easy? 2, 2, 2 . . .

At the bottom of that chart, you'll see a graphic description of **Linear Measure** (measuring length).
12 inches = 1 foot; 3 feet = 1 yard ,therefore 36 in. = 1 yard; 1 mile =5280 ft.

Weight Measure: consists of 16 ounces = 1 pound; 2000 pounds = 1 ton.

Time Measure: 60 seconds = 1 minute; 60 minutes = 1 hour; 24 hours = 1 day.

(Metric Units of Measure)
The United States is the largest major nation of the world that is not **totally** on the Metric System! We are a 2-system country—everything we buy has both Customary and Metric Measure identified. Temperature is measured in Fahrenheit (Customary) and Celsius (Metric) degrees. In this study we are going to show a very simple way to convert **within the Metric System**. Since the Metric System is based on **Powers of Ten** (remember), all we have to do is **move a decimal point to the left or right**!!

Length is measured in **Meters** Weight in **Grams** Liquid Volume in **Liters**
(The prefixes are the same for each unit . . . decimeter; decigram; deciliter)

Look at the chart below:

12.4 dekameters = _____ centimeters (just move the **decimal** 3 places to the right). Thus 12.4 dekameters =**12400** centimeters.

93 milliliters = ______ hectoliters (just move the **decimal** 5 places to the left). Thus 93 milliliters = **.00093** hectometers.

20 kilograms = ______ grams (just move the **decimal** 3 places to the right). Thus 20kilograms = **20000** grams.

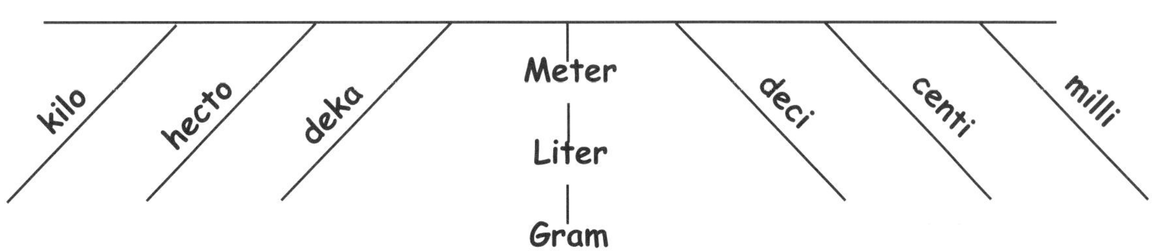

WORD PROBLEMS

The population of the world hates "word problems"!! (Math always gets blamed for this, but the culprit is "reading comprehension".) In the following problems attempt will be made to go through the thinking process.

1. Linoleum for the kitchen floor is sold by the square yard. How many square yards are needed for a floor 12 ft. by 18 ft. **and** how much will it cost at $6.50 per sq. yd.?

Find the area of the rectangular floor.
Since there are 9 sq. ft. in 1 sq. yd;
divide the number of sq. ft. by 9.
Then multiply that number by $6.50.

$A = lw$
$A = (18)(12)$
$A = 216$ sq. ft.
$216 \div 9 = 24$ sq. yd.
$24 \times 6.50 = \$156.00$

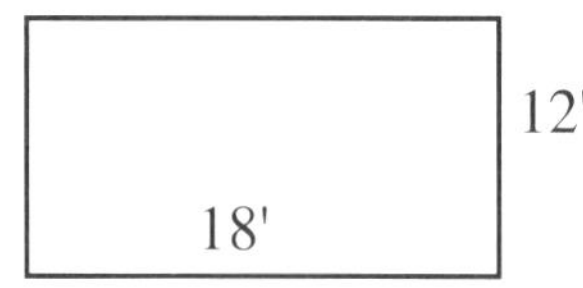

2. The original price of a shirt was $29.50, but on Saturday it could be bought for $20.00. What was the percent of **decrease** ?
First find the **decrease**. $29.50 - 20.00=$9.50
Now fill in the parts in the Standard Percent Sentence.

$$\frac{R}{100} = \frac{P}{B}$$

$$\frac{X}{100} = \frac{9.50}{29.40}$$

__X__% of _$29.50_ is _$9.50_

$$29.50X = 100(9.50)$$
$$\frac{29.50X}{29.50} = \frac{950}{29.50}$$
$$X = 32\% \text{ decrease}$$

3. Betty and her brother drink 2 quarts of milk each day. If milk costs $3.00 per gallon, what is the cost of milk for them each week?

Find how many quarts they drink each week. If 4 qts. = 1 gal. then find how many gallons they drink. Then multiply that by $3.00.

$2 \times 7 = 14$qt. per week
4 qt. = 1 gal.
$14 \div 4 = 3\frac{1}{2}$ gals. per wk.
$3\frac{1}{2} \times 3 =$
$\frac{10}{2} \times \frac{3}{1} = \frac{30}{2} = \$15$ per week

4. If a train leaves the station at 4:10 pm, what time will it arrive at the new destination after a 6 and half hour trip? 10:40 p.m.

CONTINUATION OF WORD PROBLEMS

5. A round wooden seat is cut from a 1 foot square of wood. The seat is to have a diameter of 12 in. How much wood is to be discarded?

Look at the picture! How about finding the area of the square **and** subtract (take away) the area of the circle. What is left is the discarded wood!

$A = s^2 \quad A = \pi r^2$
$A = s^2 - \pi r^2$
$A = (12)^2 - (3.14)(6)^2$
$A = 144 - (3.14)(36)$
$A = 144 - 113.04$
$A = 30.96$ sq. in.

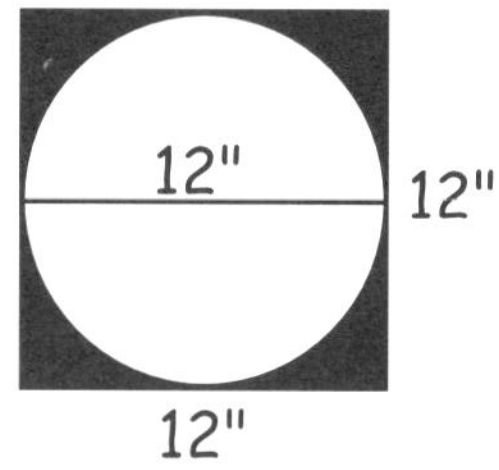

Find the Perimeter **and** the Area of the figure below.

6. This is called a "composite figure." Do you see this is a rectangle with a piece taken out? That curved piece is half a circumference of a circle. So we can find the perimeter of 3 sides + half the circumference of the circle.

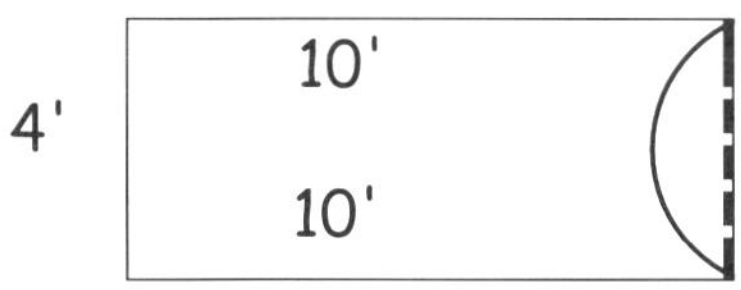

$P = 2l + w + \frac{\pi d}{2}$

$P = 2(10) + 4 + \frac{(3.14)(4)}{2}$

$P = 20 + 4 + \frac{12.56}{2}$

$P = 24 + 6.28$

$P = 30.28$ ft.

Now find the area of the figure!! If we find the area of the whole rectangle, then subtract half the area of a circle, we will have the area of this composite figure.

$A = lw - \frac{\pi r^2}{2}$

$A = (10)(4) - \frac{(3.14)(2)^2}{2}$

$A = 40 - \frac{(3.14)(4)}{2}$

$A = 40 - \frac{12.56}{2}$

$A = 40 - 6.28$

$A = 33.72$ sq. ft.